The Complete HVAC for Beginners 2024

I0427004

[4 in 1] The Updated DIY Guide to Heating, Ventilation, and Air Conditioning Systems | Residential & Commercial Solutions, Practical HVAC Tips & Tricks.

Matthew Gearhart

Table of Contents

Introduction..3

Chapter 1: Understanding HVAC Basics...............................8

Chapter 2: Residential HVAC Solutions............................. 10

Chapter 3: Commercial HVAC Solutions.............................12

Chapter 4: DIY Installation and Maintenance.....................14

Chapter 5: Advanced HVAC Tips & Tricks.......................... 16

Chapter 6: Case Studies.. 18

Chapter 7: Glossary of HVAC Terms...................................20

Chapter 8: Appendix..21

Introduction

Calling all young artists! Are you ready to unleash your creativity and bring the animal kingdom to life on paper? This exciting ebook is your passport to a world of roaring lions, prancing dolphins, and everything in between! Packed with easy-to-follow instructions and fun step-by-step guides, this book will transform you into a master artist in no time. You'll learn how to draw all your favorite animals, from cuddly cats and playful dogs to majestic elephants and soaring eagles.

So grab your crayons, pencils, and markers, and let's get started! With each page, you'll discover new skills, explore different drawing techniques, and most importantly, have tons of fun!

Introduction

"The Complete HVAC for Beginners 2024: [4 in 1]" is where you are now. The Most Up-to-Date Do-It-Yourself Guide on HVAC Systems | Commercial & Residential Solutions, Useful HVAC Tricks & Tips." The importance of a properly operating HVAC system in today's ever-changing residential and business settings cannot be emphasized. As we embark on this journey, let's explore the essential elements that define the comfort and efficiency of our living and working spaces.

Overview of Heating, Ventilation, and Air Conditioning (HVAC)

The fundamental components of building temperature management are HVAC systems, or heating, ventilation, and air conditioning. HVAC systems are essential for providing warmth in cold winter months, fresh air in well-ventilated spaces, and a refreshing breeze on hot summer days. This part offers a thorough introduction, exploring the fundamental ideas that underpin these systems and the complex equilibrium they achieve to provide a peaceful interior atmosphere.

Importance of HVAC Systems in Residential and Commercial Settings

Comprehending HVAC is not only about controlling the temperature. It includes things like energy efficiency, indoor air quality, and tenants' general well-being. A well designed HVAC system guarantees comfort, productivity, and even health in both business and residential environments. We'll look at how these techniques help design living areas that are not only more aesthetically pleasing but also more thermally managed.

Evolution of HVAC Technology

HVAC technology has advanced in a way that is really amazing. This section chronicles the development of HVAC systems throughout time, from their modest origins to their state-of-the-art advances.

Technological developments have improved these systems' efficiency while also creating new opportunities and making them smarter, more sustainable, and environment-adaptive. Readers may appreciate the advancements achieved and the possibility of further advances in HVAC technology by comprehending this progression.

Understanding HVAC Basics

We take a tour of the basic ideas that underpin HVAC systems in this part. We create the foundation for a thorough knowledge by covering everything from the fundamentals of heating to the intricate workings of air conditioning and the vital function of ventilation.

Fundamental Principles of Heating

The main component of many HVAC systems, particularly in areas with cold temperatures, is heating. We explore the fundamental ideas that underpin the production, distribution, and management of heat in homes and businesses. This part provides a foundation for understanding the nuances of generating a warm and pleasant atmosphere, including both conventional approaches and contemporary heating technology.

Ventilation Essentials

In the field of HVAC, ventilation often goes unnoticed. Effective ventilation does more than just exchange air between indoor and outdoor spaces; it also improves indoor air quality and occupant health. This section delves into the crucial function of ventilation, including subjects like air exchange rates, air filtration, and methods to maintain ideal indoor air quality.

Air Conditioning Fundamentals

The emphasis turns to the principles of air conditioning as we get into warmer areas. It's crucial to comprehend how these systems dehumidify and cool interior areas. Readers will learn about the many air conditioning unit kinds and how to create pleasant spaces even in the hottest weather, as well as the workings of refrigeration.

Components of a Comprehensive HVAC System

Familiarity with the components of HVAC systems is necessary for a comprehensive knowledge of them. The main components of a complete HVAC system—including the central unit, ducting, thermostats, and control systems—are broken down in this section. Readers may make knowledgeable judgments regarding system design, maintenance, and troubleshooting by having a solid grasp of these components.

Residential HVAC Solutions

We go beyond the fundamentals and concentrate on home HVAC solutions. Making important choices that affect day-to-day living include selecting the best air conditioner, making sure your house is properly ventilated, and designing an effective heating system. For homeowners wishing to maximize the performance of their HVAC systems, this section offers useful insights and doable advice.

Designing an Efficient Home Heating System

The correct heating system is the first step in designing a comfortable and energy-efficient house. We examine a range of heating alternatives, from contemporary radiant heating to conventional furnaces, to help readers choose the best choice for their houses. There is a thorough discussion of several factors, including environmental concerns, maintenance needs, and fuel economy.

Proper Ventilation Strategies for Residences

One of the most important components of a healthy house is ventilation. This section explores ventilation techniques designed for homes, including typical issues such exhaust from bathrooms and kitchens, as well as general airflow. Homeowners may build living environments that support well-being and improve indoor air quality by installing enough ventilation.

Selecting the Appropriate Air Conditioner for Your House

There is a dizzying number of air conditioning alternatives to choose from. Taking into account variables like room dimensions, energy efficiency, and climatic concerns, this section of the book walks readers through the process of choosing the best air conditioner. The advantages and disadvantages of every choice—from split systems to window units—are examined, enabling homeowners to make wise choices.

Troubleshooting Common Residential HVAC Issues

The most beautifully built household HVAC systems can have problems. In order to help readers solve frequent problems including inconsistent heating or cooling, unusual sounds, and thermostat malfunctions, this section provides troubleshooting techniques. Homes can guarantee the life of their HVAC systems and save maintenance costs by being aware of these problems and possible fixes.

Commercial HVAC Solutions

In terms of HVAC systems, commercial buildings have particular difficulties and specifications. The main goals of our investigation into commercial HVAC systems are regulatory compliance, energy efficiency, and scalability for big buildings.

Scaling HVAC Systems for Commercial Buildings

The size and function of commercial buildings vary greatly, necessitating the need for HVAC systems that can adapt. This section of the book looks at factors to take into account when scaling HVAC systems for commercial facilities. It covers things like zoning, load estimates, and integrating cutting-edge technology to satisfy the unique needs of huge buildings.

Energy-Efficient Solutions for Large Spaces

Energy conservation is a top priority in business environments. The techniques and innovations aimed at improving the HVAC systems' energy efficiency in vast areas are discussed in this section. With the help of sophisticated control systems and variable refrigerant flow systems, readers may optimize energy use without sacrificing comfort.

HVAC Maintenance in Commercial Environments

The durability and efficiency of commercial HVAC systems depend heavily on maintenance. Proactive maintenance strategies designed for business environments are described in this section. The need for expert service, predictive maintenance strategies, and routine inspections are all discussed, offering a road map for guaranteeing the dependability of HVAC systems in busy commercial settings.

Regulations and Compliance for Commercial HVAC

It is essential for facility managers and owners of commercial real estate to navigate the regulatory environment. A thorough summary of the laws pertaining to commercial HVAC systems is given in this section of the book. Readers get a thorough grasp of compliance requirements and the consequences of non-compliance, covering everything from environmental issues to safety regulations.

DIY Installation and Maintenance

This area is devoted to do-it-yourself HVAC installation and maintenance, giving readers the tools they need to take charge of their systems. Readers are given the information and confidence to take on their own HVAC tasks with the help of step-by-step instructions, safety precautions, and troubleshooting suggestions.

Step-by-Step Guide to DIY HVAC Installation

This chapter offers a step-by-step tutorial on installing HVAC systems in residential settings for the daring do-it-yourselfers. Readers learn about everything from ducting installation to size estimates.

Chapter 1: Understanding HVAC Basics

Heating, Ventilation, and Air Conditioning (HVAC) systems play a crucial part in the great symphony of contemporary life, where comfort and technology interact. This chapter is our manual, exploring the fundamentals of HVAC, elucidating the laws that regulate its functioning, and illuminating the vital elements that come together to provide temperature management.

Fundamental Principles of Heating

The HVAC performance begins with Heating, an energetic elemental dance. Raising a space's temperature fundamentally involves the transmission of heat. Studying the basic ideas entails investigating a range of heating techniques, from more modern radiant heating options to more conventional systems like furnaces.

The main sources of heat for many heating systems are furnaces, which burn fuel, usually oil, propane, or natural gas. Heat is released after combustion and is then transferred throughout the room via pipes or ducting. A furnace's efficiency is affected by a number of things, including insulation, maintenance procedures, and fuel type.

Conversely, radiant heating depends on the direct transmission of heat to surfaces and things within a space. This technique, which is often used in wall panels or flooring systems, provides a more pleasant and equal heat distribution. Comprehending the subtleties of these heating principles enables experts and homeowners to make well-informed decisions based on environmental, comfort, and efficiency factors.

Ventilation Essentials

In the second part of our HVAC story, ventilation takes center stage. Ventilation is more than just controlling temperature; it's about constantly exchanging air between interior and outdoor spaces. This procedure is essential for preserving the best possible indoor air quality in addition to refreshing the air.

Natural and mechanical methods are used to create ventilation. Openings like windows and vents are necessary for natural ventilation in order to allow fresh air to circulate. Conversely, mechanical ventilation regulates air circulation via the use of fans and duct systems. The decision between these approaches often comes down to variables including building design, climate, and energy efficiency objectives.

One of the most important factors in ventilation is indoor air quality. Enough ventilation is necessary to reduce indoor pollutants such as volatile organic compounds (VOCs), odors, and allergies. Having a basic understanding of ventilation principles enables building managers and homeowners to create breathable, health-promoting settings.

Air Conditioning Fundamentals

As we go forward with our HVAC performance, the scene is prepared for air conditioning, its cool counterpart. This action comprises the complex science of eliminating heat from interior areas to provide relief from oppressive heat. The refrigeration cycle is the foundational idea of air conditioning.

Fundamentally, refrigeration is the process of moving heat from one place to another. A refrigerant—typically a fluid with certain thermodynamic properties—is moved via a closed loop in an air conditioning system. Through a cycle of compression, condensation, expansion, and evaporation, the refrigerant draws heat from the inside and releases it outside.

Uncovering the mysteries of air conditioning requires an understanding of the factors involved in this refrigeration cycle, such as temperature, pressure, and phase shifts. The kind of refrigerant utilized, system architecture, and appropriate maintenance procedures are some of the variables that affect an air conditioning system's efficiency.

Components of a Comprehensive HVAC System

When our HVAC system performs at its peak, it's important to highlight the parts that make up a whole system. A well-thought-out HVAC system is a harmonious combination of several components. Comprehending these constituents endows users with the ability to make knowledgeable selections about system architecture, upkeep, and diagnosis.

The central unit is the brains of the system, whether it is an air conditioner or a furnace for heating or cooling. The building's circulatory system, or ductwork, moves warm or cooled air throughout the structure. As conductors, thermostats coordinate the system's operation according to user preferences.

Because they capture dust, allergies, and other particles, filters are essential for preserving the quality of the air within buildings. A simple yet efficient maintenance procedure is to clean or replace filters on a regular basis. Fans and exhausts are two components of ventilation systems that provide a constant supply of fresh air, improving air quality and comfort.

Chapter 2: Residential HVAC Solutions

After discussing the fundamentals of HVAC, we will now turn our attention to the residential market. Setting the groundwork for residents and homeowners, Chapter 2 offers advice on how to customize HVAC systems for happy living quarters. This chapter serves as a guide for establishing the most comfortable environment possible within the limits of your house, from constructing effective heating systems to resolving frequent problems.

Designing an Efficient Home Heating System

Developing an efficient and successful heating system is the first step in our home HVAC journey. There are several alternatives available to homeowners, and each has its own set of factors to take into account. Choosing a heating source—a heat pump, a furnace, or radiant heating—is crucial to this selection.

As the workhorses of house heating, furnaces are excellent at quickly heating up areas. They provide flexibility depending on local availability and environmental factors and come in a variety of fuel alternatives. Conversely, heat pumps have a dual function by serving as both cooling and heating systems. They are highly praised for their efficiency, especially in temperate settings. Because they distribute heat evenly, radiant heating systems are preferred for wall or floor installations that provide a comfortable atmosphere.

Appropriately sizing the heating system is an important factor. While an undersized system finds it difficult to satisfy heating needs, a huge system may result in inefficiencies and discomfort. This section gives homeowners the resources they need to do load calculations. By accounting for elements like windows, insulation, and the local temperature, they can choose a heating system that maximizes efficiency while maintaining comfort.

Proper Ventilation Strategies for Residences

Once again, ventilation is important, but this time it's in relation to homes. In addition to air exchange, ventilation is essential for controlling moisture, maintaining indoor air quality, and promoting general comfort in dwellings. This section guides homeowners through the factors to take into account when putting appropriate ventilation techniques into practice.

The potential of natural ventilation techniques, such well-placed windows and vents, to take advantage of temperature variations and prevailing winds is investigated. Exhaust fans and whole-house ventilation systems are examples of mechanical ventilation choices that are examined in detail to provide insights into their uses and advantages.

A major area of concern is indoor air quality, with advice on how to keep a healthy home. It is explained how humidity management, air purifiers, and air filters all contribute to the best possible indoor air quality. Understanding the significance of ventilation in homes allows homeowners to design environments that enhance their general well-being in addition to being comfortable.

Selecting the Appropriate Air Conditioner for Your House

Once again, ventilation is important, but this time it's in relation to homes. In addition to air exchange, ventilation is essential for controlling moisture, maintaining indoor air quality, and promoting general comfort in dwellings. This section guides homeowners through the factors to take into account when putting appropriate ventilation techniques into practice.

The potential of natural ventilation techniques, such well-placed windows and vents, to take advantage of temperature variations and prevailing winds is investigated. Exhaust fans and whole-house ventilation systems are examples of mechanical ventilation choices that are examined in detail to provide insights into their uses and advantages.

A major area of concern is indoor air quality, with advice on how to keep a healthy home. It is explained how humidity management, air purifiers, and air filters all contribute to the best possible indoor air quality. Understanding the significance of ventilation in homes allows homeowners to design environments that enhance their general well-being in addition to being comfortable.

Troubleshooting Common Residential HVAC Issues

Periodically, even the best-designed household HVAC systems have malfunctions. By the time this chapter's last act comes to a close, homeowners will know how to solve frequent problems and make sure their systems keep working well.

Common issues covered in this area include thermostat differences, unusual sounds, and uneven heating or cooling. Homeowners may reduce maintenance expenses and increase the longevity of their HVAC systems by being aware of the possible causes and remedies. Additionally, safety precautions are stressed while troubleshooting, creating a safe atmosphere for do-it-yourselfers.

Chapter 3: Commercial HVAC Solutions

The vast arena of commercial spaces is where the curtains rise, and here is where the need for accuracy in HVAC (heating, ventilation, and air conditioning) systems hits unprecedented heights. In-depth discussions of energy-efficient solutions, the need of maintenance, the challenges of scaling HVAC systems for commercial buildings, and negotiating the complicated web of rules and compliance are covered in Chapter 3.

Scaling HVAC Systems for Commercial Buildings

The HVAC performance is larger in scale in the big theaters of commercial buildings. The difficulty lies not only in ensuring comfort but also in coordinating systems to meet the many requirements of expansive areas. The complexities of scaling HVAC systems for commercial buildings are covered in this section, which offers engineers, architects, and facility managers a road map.

The main focus is on load calculations, which are essential to system design. Careful calculations are needed to establish the heating and cooling loads in commercial facilities, such as offices, malls, or industrial plants, due to their specific needs. To guarantee accurate temperature management, zoning strategies—where various locations have varying HVAC needs—are investigated.

One important factor to think about is the decision between centralized and decentralized systems. Centralized systems are efficient and simple to operate since they can oversee many zones from one location. Decentralized systems provide redundancy and flexibility since they are dispersed among several locations. Making sense of these options is made easier for decision-makers thanks to this part, which guarantees a smooth operation of the HVAC system in large commercial spaces.

Energy-Efficient Solutions for Large Spaces

Energy efficiency is often the center of attention in the financial landscape of business endeavors. This chapter's second act looks at ways to make huge rooms' HVAC systems not only pleasant, but also economically and ecologically sound.

Variable Refrigerant Flow (VRF) systems provide flexibility and energy savings by allowing the refrigerant flow to be adjusted to various indoor units. Control systems with advanced functionality, such Building Automation Systems (BAS), allow HVAC performance to be optimized and monitored centrally. Commercial buildings' carbon footprints may be decreased by integrating renewable energy sources and installing high-efficiency HVAC systems.

The significance of conducting routine energy audits is underlined. These audits find inefficiencies and provide suggestions for possible improvements. Commercial businesses that implement energy-efficient solutions not only help to preserve the environment over time, but they also save a substantial amount of money.

HVAC Maintenance in Commercial Environments

It's similar to choreographing a ballet to keep HVAC systems operating flawlessly in business settings. The complexities of HVAC maintenance that are unique to large-scale commercial settings are the subject of this chapter's act.

It is crucial to use proactive maintenance techniques, such as planned service and routine inspections. Predictive maintenance methods provide a proactive approach to system maintenance by using data and analytics to foresee possible problems. The scope and complexity of commercial HVAC systems are taken into consideration as the roles of in-house maintenance teams and external experts are examined.

In business settings, downtime may have a big financial impact. Therefore, the need of strategically planning maintenance to minimize interruptions is emphasized in this section. The chapter gives site managers a thorough overview of how to keep their HVAC systems in good working order, covering everything from replacing air filters to checking ducting and making sure cooling towers are operating at maximum efficiency.

Regulations and Compliance for Commercial HVAC

Compliance is the main focus of the commercial HVAC regulatory theater. This chapter's last act untangles the complex rules regulating HVAC systems in commercial buildings, making sure that these systems not only provide comfort but also meet safety and regulatory requirements.

Environmental rules are important, especially those pertaining to refrigerants. Examined are the phase-out of ozone-depleting compounds and the switch to ecologically friendly refrigerants. To protect people and adhere to local construction laws, safety standards—such as ventilation requirements and fire safety measures—are examined.

This section also explores the significance of record-keeping and documentation. For audits and regulatory inspections, it is essential to keep thorough records of HVAC installations, maintenance tasks, and compliance inspections. The need of being up to date on changing rules is emphasized in the chapter's conclusion in order to guarantee ongoing compliance.

Chapter 4: DIY Installation and Maintenance

As the story progresses, Chapter 4 calls homeowners and do-it-yourselfers onstage, enabling them to participate actively in the HVAC (heating, ventilation, and air conditioning) symphony. This chapter gives you the tools to execute maintenance tasks yourself and maintain maximum HVAC operation. It includes step-by-step instructions, regular maintenance practices, safety precautions, and rapid troubleshooting advice.

Step-by-Step Guide to DIY HVAC Installation

Installing HVAC systems is an ambitious first step in the do-it-yourself adventure. While expert installation is sometimes advised for intricate systems, the purpose of this part is to walk enthusiasts through the fundamental procedures of a straightforward HVAC installation.

The first phase is meticulous planning, taking into account variables including the kind of system, size, and ducting arrangement. The installation of vital parts like the central unit, ducts, and ventilation systems is covered, as well as proper insulation—a critical component for efficiency.

Safety precautions are stressed, which is a common subject in do-it-yourself HVAC projects. A safe installation procedure is ensured by knowing possible dangers, from handling refrigerants to electrical connections. With system testing and calibration, the guide finishes, giving do-it-yourselfers a thorough grasp of the installation process.

Regular Maintenance Practices for HVAC Systems

In this act, maintenance—the foundation of an HVAC system's longevity—takes center stage. Do-it-yourselfers are led through routine maintenance procedures that enhance the dependability and efficiency of their HVAC systems.

We demystify changing air filters, an easy yet important chore. Examined are the significance of cleaning condenser coils, lubricating moving components, and doing ductwork inspections. Homeowners may guarantee consistent performance and increase the longevity of their HVAC systems by implementing these principles into a periodic maintenance program.

The need of seasonal maintenance is also covered in this section, along with particular chores for the heating and cooling seasons. With a checklist in hand, do-it-yourself enthusiasts may easily traverse the ever-changing needs of HVAC maintenance throughout the year.

Safety Measures during DIY HVAC Projects

When it comes to the DIY theater, safety comes first. This statute emphasizes how crucial it is to put safety precautions in place before beginning any do-it-yourself HVAC project in order to protect homes and hobbyists.

It's critical to comprehend electrical safety, from shutting off power sources to using the right grounding measures. Refrigerant handling is a potentially dangerous operation that calls for prudence and attention to safety regulations. Personal protection equipment (PPE) is presented as a necessary tool for do-it-yourself projects.

The act also highlights how important it is to be aware of one's own limits. While do-it-yourself tasks may be rewarding, knowing when to call a professional is essential for the HVAC system's durability and safety.

Troubleshooting and Quick Fixes for DIY Enthusiasts

DIY enthusiasts may handle typical HVAC system problems by learning how to troubleshoot them in the last act of Chapter 4. Setting the setting for problem-solving and identification guarantees a smooth performance.

Many households have experienced uneven heating or cooling; possible explanations include clogged vents or unbalanced ducting. Unusual sounds, which are often a sign of deeper problems, are explained to help do-it-yourselfers identify the issue's origin.

Tips for recalibration and troubleshooting are provided for thermostat problems, which are a frequent obstacle. Homeowners may reduce downtime and even avoid expensive repairs by being aware of the possible reasons and applying simple solutions.

Chapter 5: Advanced HVAC Tips & Tricks

Chapter 5 sets the framework for advanced HVAC by using cutting-edge technology and strategic insights. This chapter reveals the emerging trends that will influence the development of heating, ventilation, and air conditioning (HVAC) systems in the future, ranging from the incorporation of smart technology to performance optimization for cost effectiveness.

Smart Technologies in HVAC

As the first act progresses, smart technologies come into focus, turning HVAC systems into sentient machines that react to environmental cues and human preferences. With its sensors and connection, smart thermostats provide owners unmatched control over the temperature inside their homes.

The masters of the smart HVAC symphony, machine learning algorithms allow systems to adjust and pick up on human behavior. The advantages of smart technology are examined in this part, including reduced energy use, improved comfort, and the practicality of remote control via smartphones and other gadgets.

Artificial Intelligence (AI) integration takes HVAC systems to even higher levels. In addition to optimizing energy use and anticipating maintenance requirements, predictive analytics and self-learning algorithms also provide a customized and adaptable interior environment.

Integrating Renewable Energy with HVAC Systems

The second act explores how HVAC systems and renewable energy sources may coexist peacefully. The story looks at how geothermal heat pumps and solar-powered heating systems are influencing how HVAC is developed in the future.

Solar heating systems lessen dependence on traditional energy sources by producing heat using sunshine. Geothermal heat pumps effectively provide heating and cooling by drawing energy from the steady temperatures found under the surface of the Earth. In order to contribute to a more sustainable and greener future, this section discusses the benefits and factors to take into account when incorporating renewable energy into HVAC systems.

Optimizing HVAC Performance for Cost Efficiency

In the third act, cost effectiveness becomes paramount, and strategic optimization plays a crucial role in striking a balance between affordability and comfort. To optimize HVAC efficiency while lowering operating costs, zoning methods, appropriate insulation, and energy-efficient designs are investigated.

Zoning plans, which are especially important in business environments, make sure that HVAC systems adjust to the unique requirements of various spaces. With the help of sophisticated control systems and programmable thermostats, homeowners can precisely adjust their HVAC settings to match occupancy patterns and energy consumption.

As discussed in previous chapters, routine maintenance procedures are essential to cost effectiveness. HVAC systems operate longer and use less energy when routine maintenance is performed, including filter replacements, inspections, and possible problem solving.

Future Trends in HVAC Technology

The focus shifts to the future in the last act of Chapter 5, where the story is shaped by anticipated developments in HVAC technology. The future looks exciting and revolutionary, with new refrigerants emerging and HVAC and smart grid technologies coming together.

The goal of creating net-zero energy buildings—those that generate as much energy as they consume—becomes the foundation. By combining smart grid technologies with HVAC systems, a future where buildings actively engage in a dynamic and responsive energy environment is envisioned.

The HVAC landscape is expected to become more intelligent and adaptable with the use of advanced sensors, IoT connection, and real-time data analytics. This section gives an overview of the advances that will change our understanding of and interactions with temperature control systems as the story of HVAC technology develops.

Chapter 6: Case Studies

Through fascinating case studies, Chapter 6 allows readers to explore practical uses for Heating, Ventilation, and Air Conditioning (HVAC) systems. This chapter provides ideas and inspiration from a variety of HVAC endeavors, ranging from inventiv commercial solutions to successful home installations and the results of do-it-yourself initiatives.

Successful Residential HVAC Installations

In the first section of Chapter 6, excellent projects that combine comfort, economy, and creativity are highlighted in relation to home HVAC systems. Every case study provides a unique viewpoint on attaining the best possible interior temperature management, ranging from modest houses to opulent mansions.

During the winter, frigid floors may be transformed into comfortable retreats with the help of innovative heating solutions like radiant floor heating systems. Modern houses with energy-efficient air conditioning systems seamlessly incorporated into them provide cool comfort during the sweltering summer months without sacrificing sustainability or beauty.

Case studies explore the complexities of system design, difficulties encountered during installation, and comfort and energy-saving results attained. Through the analysis of actual cases, homeowners may get important knowledge about recommended procedures and new developments in residential HVAC.

Innovative Commercial HVAC Solutions

In commercial settings, where HVAC systems are customized to satisfy the various demands of organizations and industries, the second act of Chapter 6 takes place. In the workplace, in retail stores, in factories, and in apartment buildings, creative HVAC solutions are essential to raising comfort and productivity.

Large-scale HVAC installations highlight the complexities of temperature management in broad commercial spaces. These installations are characterized by centralized systems and sophisticated control mechanisms. Demand response programs and the incorporation of renewable energy sources are two examples of energy management techniques that demonstrate a dedication to sustainability and economy.

Case studies illustrate the difficulties encountered and the creative solutions used to solve them. For architects, engineers, and facility managers negotiating the complexity of commercial HVAC, each case study provides insightful insights on anything from maximizing air flow in open-plan offices to guaranteeing accurate temperature regulation in data centers.

Real-Life DIY HVAC Projects and Outcomes

The spirit of do-it-yourself (DIY) enthusiasts who take up HVAC projects with enthusiasm and inventiveness is celebrated in the concluding act of Chapter 6. Do-it-yourself initiatives, ranging from basic upkeep to complex system installs, provide valuable perspectives on the real-world difficulties and satisfying results of direct HVAC engagement.

Success tales of do-it-yourself projects range widely, from installing ductless mini-split systems for customized comfort to replacing thermostats to save energy. Every project showcases homeowners' creativity and will to take charge of their interior spaces.

Case studies explore the knowledge gained, potential dangers, and successes attained from do-it-yourself HVAC projects. DIY enthusiasts encourage others to take up HVAC projects with expertise, imagination, and a sense of adventure by sharing their experiences and results.

Chapter 7: Glossary of HVAC Terms

In order to help readers understand the nuances of Heating, Ventilation, and Air Conditioning (HVAC) systems, Chapter 7 provides an extensive glossary of HVAC terms. This dictionary provides readers with the language needed to understand the wide range of HVAC discourse, from basic ideas to specialist jargon.

Comprehensive List of HVAC Terminology

Chapter 7 begins with an exhaustive glossary of HVAC terms that have been carefully selected to cover the range and depth of ideas in the subject. Readers meet terminology from A to Z that range from fundamental ideas to sophisticated methods, each providing insight into the complex world of HVAC.

HVAC system comprehension begins with basic ideas about heating, ventilation, and air conditioning. Compressor, evaporator, and condenser terminology, among other terms pertaining to system components, clarifies the inner workings of HVAC technology.

A deeper exploration of the science and engineering concepts behind HVAC design and operation is provided by advanced subjects including psychrometrics, load calculations, and refrigerant qualities. By being acquainted with the jargon, readers are better able to understand the intricacies involved in temperature control systems.

Definitions and Explanations for Clarity

Chapter 7's second act provides explanations and contextual insights to improve comprehension, going beyond simple definitions. Every phrase has a brief definition that explains its meaning and use in the context of HVAC.

To close the gap between theory and practice, examples and real-world applications are added to the definitions. Technical jargon is demystified and complex ideas are reduced to provide readers a clear knowledge of HVAC terms.

Additionally, a comprehensive understanding of linked ideas within the HVAC domain is provided via cross-references and linkages to relevant terminology. Readers get a deeper understanding and mastery of HVAC concepts by delving into the subtleties and relationships between terms.

Chapter 8: Appendix

Chapter 8 provides an extensive list of supplementary materials, tools, references, and frequently asked questions (FAQs) to enhance the reader's exploration of the HVAC (heating, ventilation, and air conditioning) industry. This appendix provides useful resources for further research, enhancing the reader's comprehension and offering a wealth of additional information, methods, and perspectives.

Additional Resources and References

Act I of Chapter 8 provides readers with an abundance of further materials and references, building on the basic information provided in earlier chapters. To enhance their comprehension of HVAC concepts and practices, readers may access a plethora of material via books, articles, online forums, and industry publications.

Reputable books written by professionals in the field are invaluable resources for further research and learning. Scholarly articles and research papers provide valuable perspectives on the latest advancements and developing patterns in HVAC technology and innovation.

Within the HVAC community, online tools like blogs, websites, and forums provide a place for continuous education, conversation, and cooperation. Through utilizing these supplementary materials and sources, readers set out on an ongoing path of learning and development in the HVAC industry.

Recommended Tools for DIY HVAC Projects

Act II of Chapter 8 presents a carefully chosen list of suggested resources that are necessary for do-it-yourselfers starting HVAC projects. Readers are armed with the tools they need to handle installation, maintenance, and troubleshooting chores confidently and effectively, ranging from simple hand tools to specialist equipment.

Any DIY toolbox must start with basic hand tools like pliers, wrenches, and screwdrivers since they make everyday chores easier to complete. Accurate diagnostics and adjustments are made easier by measuring devices like multimeters, thermometers, and pressure gauges.

Vacuum pumps, refrigerant recovery units, and leak detectors are examples of specialized equipment that can handle more complex HVAC projects and troubleshooting situations. The use of protective gear, such as respirators, safety glasses, and gloves, guarantees the wellbeing of do-it-yourselfers throughout their projects.

DIY enthusiasts may improve their talents and produce HVAC projects of professional quality by investing in products that are suggested for their particular demands and ability level.

Frequently Asked Questions (FAQs)

Chapter 8's last section answers frequently asked questions (FAQs) that come up in the HVAC industry. This area offers thoughtful responses and recommendations to frequently asked questions and concerns, based on the queries and worries of the readership as a whole.

FAQs include a broad variety of subjects, such as choosing the right system, installing it, maintaining it properly, diagnosing issues, and energy efficiency factors. Every question is answered in a straightforward and succinct manner, providing readers with professional insights and workable answers to help them in their HVAC endeavors.

FAQs are also a great tool for resolving any concerns and diagnosing typical problems that may arise over the course of an HVAC system's lifetime. Through the use of the combined knowledge extracted from these answers, readers are able to successfully complete their HVAC projects and overcome obstacles with assurance.

www.ingramcontent.com/pod-product-compliance
Lightning Source LLC
Chambersburg PA
CBHW071020290526
45795CB00005B/1876